The Book of Hive

Strategy, Tips and Tactics

by

Steve Dee

First published in 2015
as *Hive: The Boardless Board Game*

This edition published 2020

Copyright © Steve Dee 2020

The right of Steve Dee to be identified as the author of this book has been asserted in accordance with Section 77 and 78 of the Copyrights, Designs and Patents Act 1988. Copying of this manuscript, in whole or in part, without the written permission of the author or his publisher is strictly prohibited and would constitute a breach of copyright held.

To contact the author please visit www.skdinning.com

Contents

Introduction 5

Chapter 1: The Rules 11

Chapter 2: Basic Tactics 25

Chapter 3: Mistakes to Avoid 42

Chapter 4: The Opening 47

Chapter 5: More Advanced Tactics 56

Chapter 6: The Count 69

Chapter 7: Beetlemania 73

Chapter 8: Interview 79

Chapter 9: Final Thoughts 94

Introduction

I know that most of you will skip this section.

The 'Introduction' in a book like this is usually where the author goes on at great length about himself, and tries to convince you that he is qualified to give you his advice. I always skip these sections myself, so I do not blame you.

Go on then. Skip ahead. I will never even know.

If you are still with me, I want to tell you my story. It is the story of a young Englishman's journey of personal growth and triumph over adversity. The scene is the 2014 *Hive* World Championship, in the glamorous setting of the Stuttgart Hilton Hotel, Germany.

Our hero enters the venue, the hotel ballroom, and looks out across a sea of people. The usual partying that the ballroom sees is absent today. Although people are paired off, no-one dancing; there is no music, only fierce concentration.

Our hero takes a deep breath, and …

No … I'm sorry … I have to stop there.

I cannot do this.

I have never really been to the *Hive* World Championship. I don't think they even have one, (maybe online?). And I don't know why I chose Stuttgart as the venue for this fictional event. I guess I thought that it sounded plausible because, as you might know if you are into modern board games, Germany rather leads the world in pursuit of that particular hobby, having invented so many of the modern classics like *Carcassonne* and *Settlers of Catan*.

Hive however, is actually a British game. It was invented by an Englishman called John Yianni. He lives in a place called 'Potters Bar' which sounds like a pub for snooker players but is actually a town just north of London. It has also been the home of Martin Freeman, Acker Bilk, and the guitarist from Suede. For me this conjures up a lovely image of the residents of the town sitting around in hobbit-holes, playing *Hive*, with a clarinet version of *Animal Nitrate* playing in the background, which I think sounds like the perfect night in.

I am actually an Englishman too. That bit in my fantasy was true (although if I am honest "young Englishman" is pushing it a bit). I should mention though that I have never actually met John Yianni. If you are an American reading this, you probably think of England as being a small country, but it is not so small that we all know each other.

I first came across *Hive* when I saw it recommended on a board game review website called *Shut Up and Sit Down* (a marvellous site by the way). I was intrigued when one of the writers, Quinns, remarked on how many awards it had won, including Mensa game of the year, and I smiled when he said it was the only board game you could play underwater. It was, however, his comment that playing it was "creating a tiny arena of minds" that made me decide to buy it.

Since then I have been playing it every chance I can, in lunch breaks at work, online against strangers, and at home. Most of my playing at home has been against myself, moving the pieces around trying to work out the best strategy. My eleven-year-old daughter thinks this is "sad". However, she also thinks that the definitive version of the John Lennon song *Imagine* is the one by Connie Talbot, the seven-year-old who came second in *Britain's Got Talent*. So what does she know?

For the past couple of years I have immersed myself in the world of the *Hive* – a place where Bees, Spiders, Ladybugs, Ants, Pill Bugs, Grasshoppers and Beetles all live together, in a hive. I am not quite sure why Ladybugs and Ants are living in a hive, or indeed where all the other bees have gone. I am also not quite sure why the spiders do not just eat the ants. I do not know why there is so much interspecies co-operation in the hive, and I am not sure whether the creatures' desire to be next to the Queen of the opposite colour is a sign of racial tension or racial harmony. I do not care either.

By the way, as an Englishman, I want you to know how uncomfortable I feel saying the word 'Ladybug'. On my side of the Atlantic we say 'Ladybird'. I do not know why or how we have ended up with different words for the same creature. How can I find out who decides these things? I don't know whether to ask an etymologist or an entomologist.

(That, by the way, is my favourite joke in this book. I hope you enjoyed it as much as I did.)

I suspect that the English originally invented the word and the Americans changed it as soon as they got the chance. Perhaps that was part of the reason the Pilgrim Fathers left England in the first place? Perhaps what they really wanted was the freedom to practice religion in their own way, and the freedom to stop using misleading entomological nomenclature ("They're bugs not birds dammit!"). I guess I will never know.

Sorry, I wandered off the subject a bit there. I was supposed to be trying to convince you that I know what I am talking about. I might not be a world champion *Hive* player, but I was a schoolboy regional champion chess player (Yes, schoolboy. Shut up. It still counts.) and I am now a statistical analyst by profession; so I do know about strategy games and about analysis. I have also earned the rank 'Expert Player' for when I play *Hive* online. You can trust me.

As a fan of the game, I am aware that there is a shortage of books aimed at beginners and intermediate players. So, I wrote one. This one, in fact. I hope you enjoy it.

Chapter 1: The Rules

What is Hive?

It sounds like some horrid illness, some terrible disease.

"I've got Hive Pocket!" *you shout down the telephone to your GP, sweaty hand gripping the receiver.*

~ Paul Dean (Shut Up & Sit Down)

You can skip this chapter too if you already know how to play.

Perhaps you do not know, though.

Perhaps, like I did, you bought a German edition of the game and were disappointed to find that the rules do not have an English translation.

Here then, for beginners (or for reference for experienced players), are the rules of the game.

Setup

Each player takes all the pieces of one colour and places them face up in front of them, as their supply. The pieces are hexagonal tiles representing insects* and in a standard set, each player will have 11 pieces as follows:

- 1 Queen Bee
- 2 Spiders
- 2 Beetles
- 3 Grasshoppers
- 3 Ants

Some versions of the game have optional extra pieces, called 'expansion' pieces, as follows:

- 1 Mosquito
- 1 Ladybug
- 1 Pill bug

*- *okay, insects, arachnids, and whatever the hell a pill bug is.*

Overview

The object of the game is to be the first to surround your opponent's Queen Bee on all six sides. The pieces surrounding the Queen Bee can be a mixture of both your pieces and your opponent's.

Players take turns, and on each turn, a player may do one of two things - place a piece, or move a piece.

The official rules do not specify which player goes first, and we usually follow the rules that apply for chess, which is that we randomly determine who is white, and then white goes first.

For the examples in the rest of this book, we will assume that 'player one' is white.

Placing a Piece

A player can choose a new piece from his supply and place it in play. It must be alongside at least one of that player's existing pieces and it cannot be alongside any of the opposing player's pieces.

The only exception to this rule is the first piece played by each player. The very first piece placed is of course, on its own. The other player's first piece is placed adjacent to the first player's first piece.

Each player must place the Queen Bee within his first four turns.

Moving a Piece

Once a player's Queen Bee has been placed (but not before), that player may choose to move an existing piece instead of placing a new one. When moving a piece, the rule about not touching an opponent's piece no longer applies.

Each of the insects (yes, I am sticking with insects; leave me alone) moves differently.

Queen Bee

The Queen Bee has very limited movement; she can only move one space at a time around the perimeter of the hive.

The Queen Bee also follows the 'Freedom to Move Rule'. This rule will be explained properly later on, but for now let us say that pieces that follow the Freedom to Move rule have to be able to slide into their new spaces without having to move other pieces out of the way.

Ant

The Ant may move only around the edge of the hive, like the Queen Bee. Unlike the Queen Bee though, it may move as many spaces as the player wishes.

The Ant also follows the Freedom to Move Rule.

Spider

Like the Queen Bee and the Ant, the Spider can move around the perimeter of the hive. It must however move exactly three spaces - no more, no less - in a direct path.

A movement of one 'space' equates to a shift to a different empty hex shaped area that is adjacent to both the current space and to one other piece.

The Spider also follows the Freedom to Move Rule.

Beetle

The Beetle can move only one space at a time, just like the Queen Bee. However, unlike the Queen Bee, a Beetle may also climb on top of any adjacent piece, and then if the player so wishes can move one space at a time over the top of the hive, or drop down into any adjacent empty space.

Beetles can climb on top of other Beetles even when they are on top of another piece or pieces.

A piece with a Beetle on top of it cannot move, and its owner may not place new pieces adjacent to it. Instead, the owner of the top Beetle in the stack may place pieces of their colour adjacent, following the usual rules.

Beetles do follow the Freedom to Move Rule when moving around rather than climbing or dropping. Note that this applies even when on top of the hive; if there are three or more beetles up there it is possible (though rare) that a beetle might not be able to slide into the place that its owner wants it to go to.

Grasshopper

The Grasshopper is, as one might expect, a jumping piece. It moves by jumping over any number of other pieces (but not empty spaces) in a straight line to the first unoccupied space it comes to. When it jumps, it must lead with an edge, not a corner, so that it jumps over a single row of pieces.

Ladybug

The Ladybug moves exactly three spaces. The first move is up onto the top of the hive; the second is on top of the hive; the third is to move back down again. This means that the Ladybug, like the Grasshopper, can move into or out of surrounded spaces.

Pill bug

The Pill Bug moves like the Queen Bee – one space at a time – but it also has a special ability that it may use <u>instead</u> of moving. This ability is that the Pill Bug may move an adjacent piece (on its own side or the opponent's) two spaces: one up onto the Pill Bug itself, then another down into an adjacent empty space.

There are some restrictions:

- The Pill Bug may not move the piece most recently moved (either directly or by another Pill Bug) by the opponent.
- The Pill Bug may not move any piece in a stack of pieces.
- The Pill Bug may not move a piece if it violates the One Hive Rule by splitting the hive.
- The Pill Bug may not move a piece through a too-narrow gap of stacked pieces (this would violate the Freedom to Move Rule on top of the Hive – see later).
- If a Pill Bug is moved by an opponent's Pill Bug, it cannot use its special ability next turn.

Mosquito

The Mosquito takes on the movement ability of any piece of either colour that it is touching at the start of its move (a stack with a Beetle on top counts as a Beetle for this purpose).

There are a couple of exceptions to the rule:

- If a Mosquito is moved as a Beetle onto the top of the hive, then it continues to move as a Beetle until it climbs back down.
- If a Mosquito is only touching another Mosquito at the start of its move, it may not move at all.

Note that the mosquito can mimic either the movement or special ability of the Pill Bug, even when the Pill Bug it is touching has been rendered immobile by being moved by another Pill Bug.

One Hive Rule

The pieces in play must be linked at all times. A piece may never be moved such that during or after its movement, there are two separate groups of pieces in play. For example, in the diagram below, the spider cannot be moved:

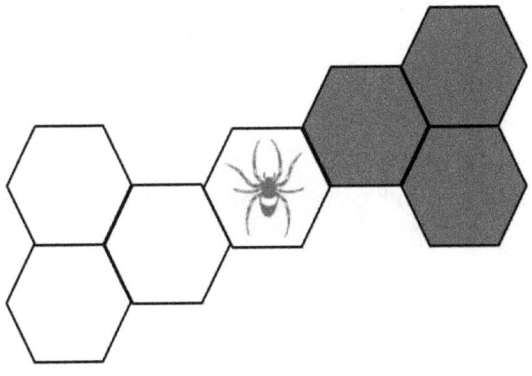

Even if as a result of the piece's move, the layout remains one group, if the hive becomes disconnected while the piece is in transit the move is not allowed. For example, the move below is illegal.

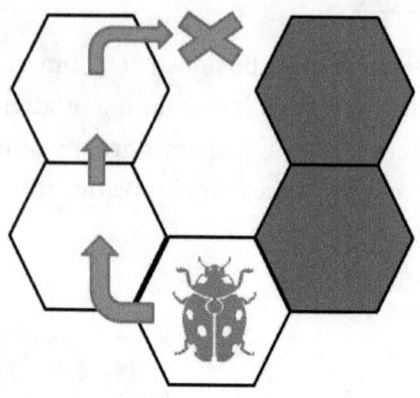

By the way, you have probably noticed by now that in my diagrams I do not always identify the piece, sometimes just the colour. I do this when it does not matter what the piece is. I do it for two reasons:

1. I want to draw attention to one or more particular pieces and do not want any distractions.
2. It makes drawing the diagrams easier.

Freedom to Move Rule

Pieces which move around the perimeter of the hive (i.e. the Spider, the Ant, the Queen Bee, and the Pill Bug) may never move into or out of a space that is partially or completely surrounded if it can no longer physically slide into or out of that position.

The rule is that if a piece cannot be slid into or out of the space without moving another piece out of the way first, then it is an illegal move. For example, in the diagram below, the ant cannot move.

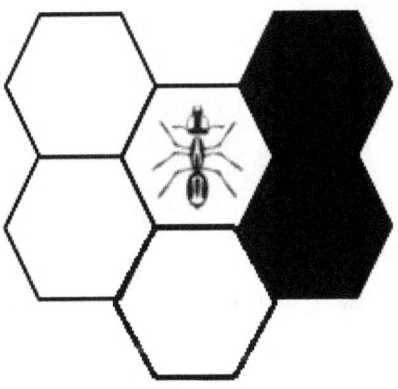

Note that this rule also applies when pieces move around on top of the hive, including when the Pill Bug moves them.

However, it does only apply to moving pieces, not pieces being placed for the first time.

Unable to Move

If one player cannot make any legal moves then they miss their turn, and the other player takes another turn.

End of the Game

The game ends when a Queen Bee is captured by being surrounded on all 6 sides by either player's pieces. The player whose Queen Bee is surrounded loses the game.

It is possible for the game to be a draw if a move results in the simultaneous surrounding of both Queen Bees.

A draw may also be agreed if each player's best move for a turn leads to an endless cycle of repetition of a series of moves. This is known as a stalemate.

Chapter 2: Basic Tactics

"Big bees have little bees upon their backs to bite them, and ... no, wait ... those little ones are actually beetles!

And also, what are all those grasshoppers doing in my beehive?"

~ Jonathan Swift, 1733

The Opening

The Queen Bee must be placed in one of the first four turns, and until a player's Queen Bee has been placed, that player may not move any pieces. It is generally preferable to place the Queen Bee before being forced to do so on the fourth turn.

A common opening is to play a spider first, then the Queen Bee in the second move, followed by an Ant, in a ^ shape, as shown below.

Alternatively, some players like to place the Queen Bee in the first move. However, if both players do this, the game is more likely to end in a draw as the Queen Bees would share three of their six surrounding hexes.

We will talk more about Openings in a later chapter.

Placing Your Pieces

Pieces should not just be placed randomly. You should always select the right piece and the right place to put it.

It is usually a good idea to bring Ants into play early in the game. You will want to use them to pin your opponent's pieces; try not to let them get pinned themselves though.

Also, if you are planning to use them, Beetles are often best brought in early as it takes time to move them and if you leave it too late they might never reach their destination.

It is not usually a good idea to play Grasshoppers early, as their movement is limited at the beginning of the game, and their special ability of being able to jump into tight spaces is more likely to be useful later in the game.

Here are some things to think about when placing your pieces:

- When you are placing an Ant, do not place it somewhere it can be pinned by a Spider or Grasshopper.
- With Mosquitos and Pill Bugs, consider their special powers before you place them.

- When you are placing a Grasshopper or Spider, put it somewhere where it can usefully go on its next turn.
- Placing a piece adjacent to your own Queen Bee is very risky. You might be thinking to yourself that you will move it later in the game. Perhaps it is a Grasshopper and you might think to yourself that it can hop out later on in a dramatic game-saving move. However, if your opponent pins the Grasshopper it could be stuck there. If this happens, you have wasted a turn, lost a piece and your Queen Bee will be partially surrounded while your opponent's pinning piece is still free to move and could later play a game-winning final move (note that as a corollary to that, if your opponent plays a piece next to his Queen Bee, it is usually a good idea to pin it if you can).

Which Pieces to Move

Some pieces are more powerful than others. If you can pin one of your opponent's more powerful pieces with one of your own weaker ones, this gives you a potential advantage because later in the game his free pieces are likely to be weaker than yours.

Ants and Mosquitos are generally considered to be the most powerful attacking pieces, though Beetles can also be very effective if used properly.

The Pill Bug is the most powerful defensive piece.

Grasshoppers have a useful ability to hop in and out of tight spaces, as do Ladybugs (very useful towards the end of the game).

The Spider however, is usually regarded as being the weakest piece because its move is so limited.

Early in the game, when the hive is small, the Spider's limitations do not affect its performance so much. For this reason, many players recommend playing a Spider early on and using it to pin the opponent's Queen Bee, Ant, or Beetle.

Pinning Opponent's Pieces

The 'One Hive Rule' states that the hive must be connected at all times. Therefore if you move one of your pieces so that it is adjacent to one of your opponent's pieces (and no other pieces), then the opponent's piece will not be able to move away without leaving your piece isolated, breaking the One Hive Rule. In this case we would say that the opponent's piece is 'pinned'.

It is usually a good idea to pin your opponent's Queen Bee early in the game, because it is very difficult to surround a piece that can move around.

In the diagram below, White has just moved his Ant to pin Black's Queen Bee.

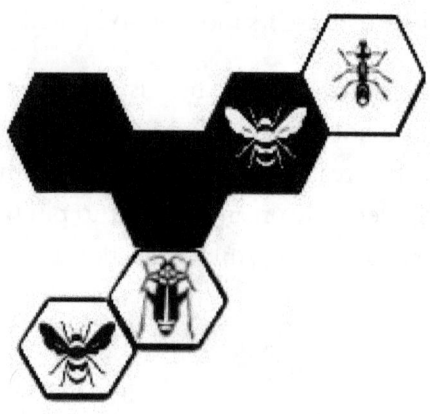

Use of the Elbow

When three or more pieces are connected in a line, but the line is not straight, and has a kink in it, that kink is known as an 'elbow'.

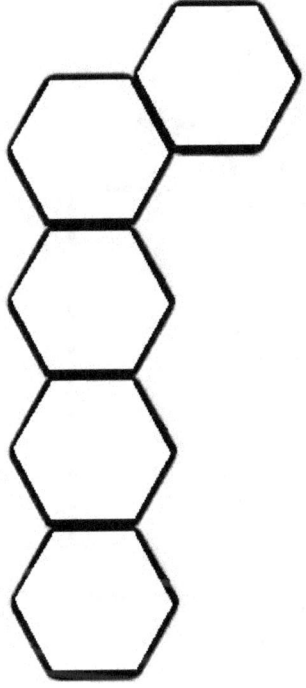

It does not matter what colour any of the pieces are.

An elbow can often be used to free a pinned (more useful) piece.

In the diagram below, White can move the Spider into the elbow, allowing the Ant to escape.

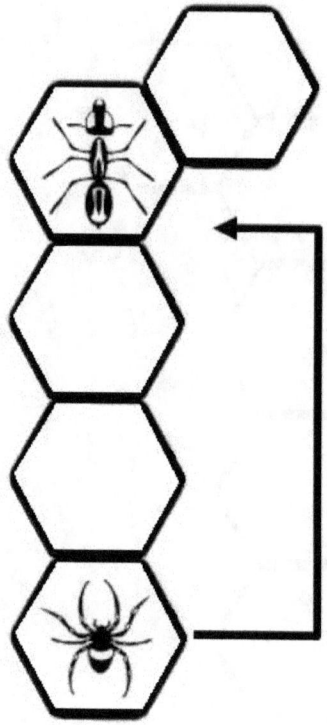

When you are the player doing the pinning, therefore, it is usually best (other things being equal) to pin in a straight line rather than create an elbow, as straight-line pins are harder to escape from.

Blocking and Circling

An alternative to pinning is 'blocking' an opponent's piece. This means surrounding it, or partially surrounding it, so that it cannot slide out, and cannot move without breaking the Freedom to Move Rule. Grasshoppers, Beetles and Ladybugs cannot be blocked.

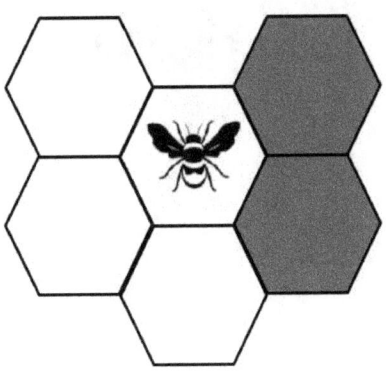

'Circling' is a variation on blocking.

Creating an incomplete circle (not a complete one or some of the opponent's pieces might be freed up to move away!) next to a Queen makes it more difficult to surround, because opponents' pieces cannot slide into one of the spaces, as illustrated below:

The gap that pieces cannot slide through is often referred to as a 'gate'.

The Pocket

When four pieces (of either colour) form a U-shape this is known as a pocket.

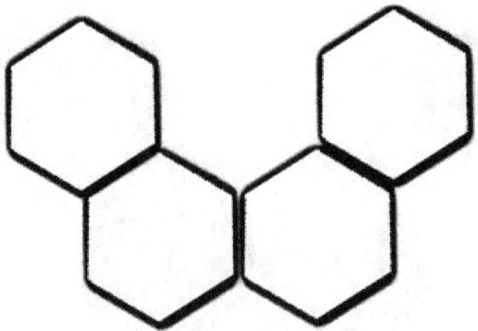

If, in addition, there is a piece in the middle, then it is a closed pocket.

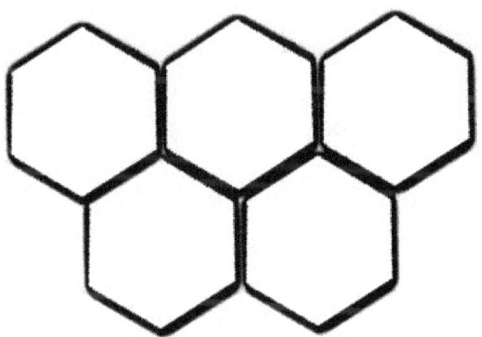

Forming a pocket around a Queen Bee will often allow it to escape (perhaps negating the effect of an earlier pin).

This is desirable if it is your own Queen Bee, but it is something you should be careful not to allow with your opponent's. If you have spent several turns surrounding your opponent's Queen Bee only for it to casually wander off to a spot where it is not surrounded at all, the game will probably be over for you.

A useful tactic is to fill a pocket in order to free up other pieces to move.

In the diagram below, white can move a Grasshopper into the open pocket, freeing the Ant, the Beetle, or the Mosquito.

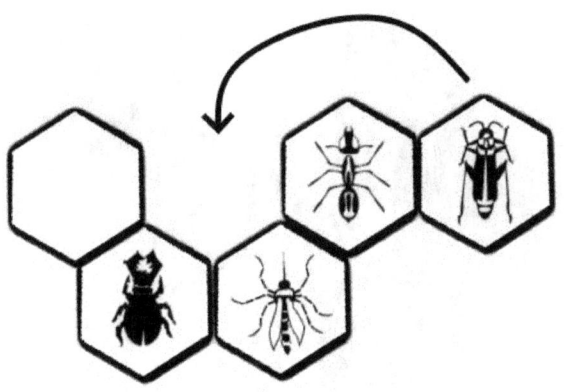

Be careful though - if any of the pieces in the U formation belong to your opponent, filling the pocket could potentially free them instead. And since your opponent will have their next move before yours, you might end up helping them instead of yourself.

Perhaps the most common occurrence of the pocket is shown in the diagram below. If both players have followed the standard opening (see beginning of Chapter 2) and Black is attacking White, something similar to the position below is likely to arise.

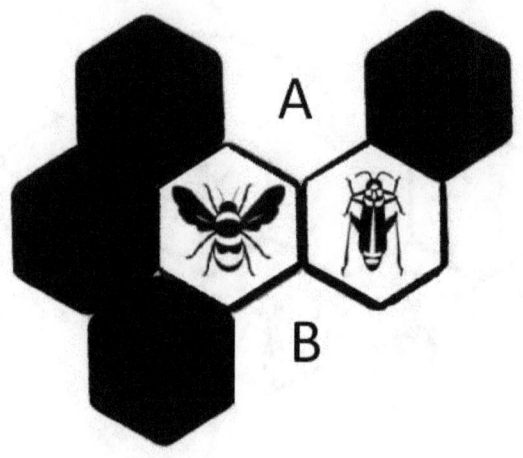

Black is close to surrounding White's Queen Bee and has, say, a free Ant (not shown) with which to attack. The Ant should be placed in position B, not A.

Position A is in a pocket, and if the Ant does go there, White's Grasshopper can move away, which is bad for Black. Filling position A should be the final (game winning) move.

When to Attack and When to Defend

It is very common for a game of *Hive* to become a race to see who can surround the opponent's Queen Bee first.

It is important to know when the race is one you are going to win, or lose. In most games, if you are behind, you will not be able to catch up. In these circumstances you must abandon the attack and defend instead. If you defend successfully, you should find that your opponent loses momentum and runs out of ways to attack you. The game then reaches a turning point, and it is now time for you to go on the offensive.

At the beginning of the game the player who goes first is ahead, so they should attack to make the most of their advantage, and attack first.

Being able to identify whether you are ahead or behind later in the game is a crucial skill, and will be discussed further in chapter four.

When and What to Place

This is a tricky question, but an important one.

How many pieces do you need in play at any one time?

Well, it often depends on whether you are attacking or defending. An attacking player will want to get their pieces out as quickly as possible, and often get into an initial rhythm of placing and moving for quite a few turns. The defending player might not be able to get as many pieces out early on, as they must react to the attacks.

In fact, the number of pieces you have in play is not actually as important as the number of mobile pieces you have in play. If all of your pieces are pinned or blocked (or pinning or blocking) and cannot move, then you will need to add some more. You should plan to have enough mobile pieces to attack, or to deal with a potential attack.

Generally speaking, it is important for both players to get some powerful pieces out early. Ants are especially useful, defensively as well as in attack. Early is often a good time for an attacker to get a spider out, when it can perform its attack with no chance of being stopped.

Regarding the pieces that can move into blocked spaces, whether by jumping in (Grasshoppers) or climbing in (Beetles and Ladybugs), it is usually a good idea to keep some of those in reserve for later in the game. These pieces might be the only way to get into certain spaces, and if they are all pinned or in the wrong location, you might find yourself with no way to win the game.

Chapter 3: Mistakes to Avoid

There was a young man from St Ive
Who bought (so he thought) the game Hive.
He thought it was funny
The box smelled of honey,
Then realised the bees were alive.

~ *Paul Perro*

If you and your opponent are both beginners, or even if you are both intermediate players, the best way for you to win is simply to make fewer mistakes than your opponent.

If you have ever played chess, you will know that most games between novices tend to be decided not by who plays the best opening, or the best long term strategy, but by who does <u>not</u> accidentally places their Queen in the path of an opponent's rook or bishop. It is the same with *Hive*. You can have a great strategy in mind, but if you make an elementary blunder, you are likely to lose the game quickly.

You should always strive therefore, to be the player who makes no elementary mistakes. The following is a list of the most common mistakes that you should try to avoid. Some have been mentioned before but I have put them all here together to form a useful reference.

Placing the Ant as an Opening Move
Don't do it. The Ant is a powerful piece and it will be blocked in.

Delaying Placing the Queen Until the Fourth Move
If you do this you will find yourself in an unfavourable position. Your opponent will be able make an effective pin, and dictate where you can and cannot place your Queen Bee.

Not Getting Enough Powerful Pieces in Play
Make sure you get some Ants out early to attack or defend with.

Not Trapping Opponent's Queen Bee
If you start to surround a Queen Bee and it moves away, you will have wasted a lot of time and resources.

Pinning a Weak Piece with a More Powerful One
Do not pin, say, a Spider with an Ant, unless you have to. Otherwise you fill find yourself out-gunned later in the game.

Self Pinning
Be careful not to pin your own pieces. If you place a piece that pins one of your more powerful pieces, even if you plan to move it next turn, your opponent could pin the new piece and he will have double-pinned you.

Accidentally Freeing a Pinned Piece
When you position a piece, be careful that you are not completing a connection somewhere that allows a previously pinned piece to move. This is especially important with the Queen Bee.

Forgetting About the Last Move
A piece which is pinning another piece cannot usually be moved for fear of freeing that piece. However, it is often overlooked that the pinning piece can be used to make the game winning final move, as it does not matter about freeing an opponent's piece then.

Placing Friendly Pieces next to Queen Bee
Your opponent could pin them, and you will have helped them to surround your own Queen Bee. Sometimes it is a good idea to place pieces there that can jump or climb out later, but try to avoid placing Spiders or Ants there.

Misplacing a Spider
They move exactly three spaces so work out where you ultimately want it to go, and count back three spaces. Always double-check it.

Wasting a Turn
Moving a piece for no reason or placing a piece that has no value is a waste of a turn As *Hive* is often a race, this could cost you the game.

Not Attacking when the Time is Right
If you miss the opportunity to attack you will not win.

Not Defending the Queen
If you know you are behind and your opponent attacks, you must defend.

Allowing a Stalemate
Do not become complacent when you are far ahead, because a decent opponent can often force a stalemate. Even if they have no chance of winning themselves, they can get a draw by defending their Queen Bee and pinning and blocking your pieces in such a way that you can never complete the victory.

Chapter 4: The Opening

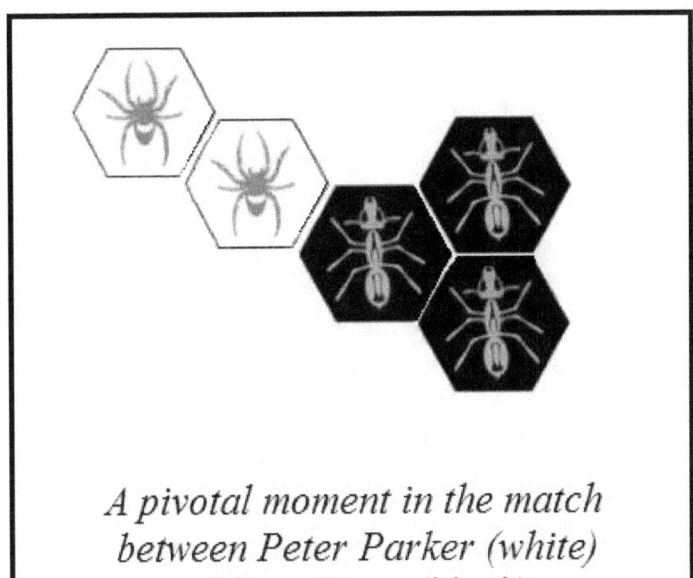

A pivotal moment in the match between Peter Parker (white) and Scott Lang (black).

Playing the Queen Bee in the first move often leads to draws, and playing it in the fourth move is recommended against, as it tends to lead to a weak position. For these reasons, most players tend to place the Queen Bee in their second or third move.

It is often stated that placing the Queen Bee in the second move means that you can move your pieces around quicker, but this is not really true if you think about it – there is nothing useful to move in the third turn; you would need to place another piece first. So, as there is no advantage in placing the Queen Bee second, I usually place mine in my third move. By placing it in my third move I can keep my opponent guessing for slightly longer.

I usually play my first three pieces in a ^ formation with my first piece at the point.

I have seen players who play their first three pieces adjacent to each other in a triangle, and this usually works out fine as one of the pieces usually gets moved away pretty quickly anyway.

However, I would caution you against playing your opening pieces in a straight line, as your opponent will pin them all at once, and you will not be able to move any of them at all.

So, I will assume we are going to place our first three pieces in a ^, with the first piece at the point. But which three pieces should they be?

1.Spider – 2.Ant – 3.Queen Bee

This is one of the openings recommended by the game's publishers. The reason why they suggest placing the Spider first, it that the first piece placed often gets trapped and does not actually play a part in the game. The Spider is the weakest piece, so playing it first is essentially a sacrifice – it allows you to save your better pieces for later.

When I first started playing, I used to lead with the Spider a lot. However, I rarely do so now. The reason for this is that I find, especially when I play with expansion pieces, that I rarely actually run out of pieces. I play quite aggressively, and the game is often over before all the pieces are placed; I almost always find that I have an unused Grasshopper left over. Therefore, there is no need to 'sacrifice' a Spider – why not lead with something potentially useful instead?

Nowadays, I usually lead with a piece that, later in the game, might be able to jump or climb out, potentially saving my own Queen Bee from being surrounded, and perhaps even surrounding the opponent's Queen Bee as well.

1.Ladybug – 2.Ant – 3.Queen Bee

For the reasons described above, I find this a very powerful opening. Playing a grasshopper first can be effective, but the Ladybug is even more of a threat in the endgame. I have won many games with the Ladybug escaping and attacking my opponent's Queen Bee.

If playing black, a good defensive piece to place fourth is the Mosquito. If you place this at the back of your V formation, it is adjacent to all three of your pieces. This means you can use it as an Ant, or later in the game it can climb out like a Ladybug and attack your opponent's Queen Bee. Also, if your Queen Bee is attacked by a Beetle, the Mosquito will turn into Beetle itself and attack back.

<u>1.Queen Bee – 2.Spider – 3.Spider</u>

This is the other opening recommended by the publishers. The way to play it is to use your Spider to pin your opponent's Queen Bee or other powerful piece on your third turn. This weakens their attack and can limit where they can place their pieces in the future. It is a very fast attack and Black can use it to seize the initiative from White.

Be careful though - a decent opponent might try to outmanoeuvre your Spiders by placing pieces in its way that make the Spider's three space move less effective. I usually prefer to substitute the second Spider with an Ant for this reason.

<u>What Not To Do</u>

Never play an Ant first.

Don't make me tell you again.

Next Few Moves

The two or three moves which follow the placement of the Queen Bee can set the tone for the whole game. You will want to adjust your strategy to respond to whatever it is that your opponent does.

Here then, are a few things to look out for.

If the Queen Bees are placed facing each other on the same side of opening pieces, so that the pieces form a C shape, as shown below:

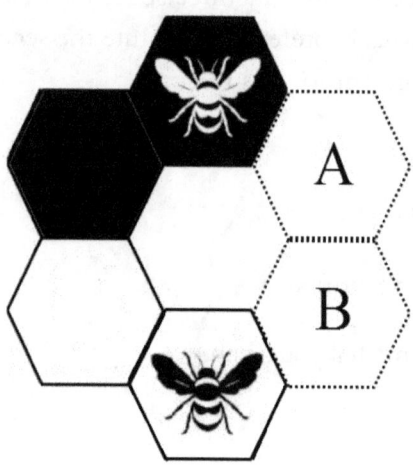

...then both players should be looking carefully at spaces A and B. White should be attempting to fill space A before Black fills space B, and vice versa.

Whoever wins that particular race (and it should be White, as they are a move ahead) will have a clear advantage, because when the loser of the race fills up the other space, they complete a ring and will most likely free up the opponent's pieces to move away.

(Despite this, games that begin in this way often end in draws because the Queen Bees share a common adjacent space.)

If the opening is one where the Queen Bees are placed on opposite sides of the opening pieces, as shown below:

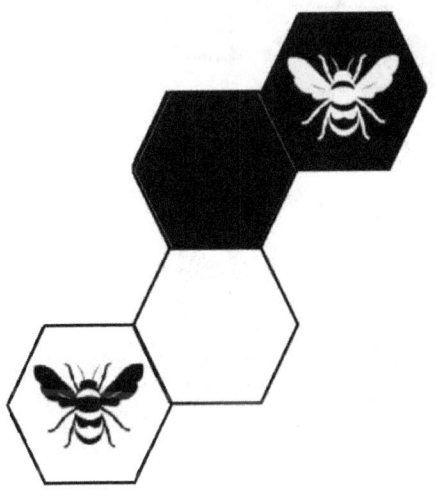

...then the game tends to be a lot more open. It is a good idea to get Ants and Mosquitos out early in this type of game, as they will be very useful for attacking and defending.

It is worth noting that it is Black who decides which of the two patterns above are played, because Black usually plays their Queen Bee after White does.

This brings us to the third most common opening pattern, where White plays their Queen Bee behind the opening piece, and Black plays theirs off to the side, as shown below:

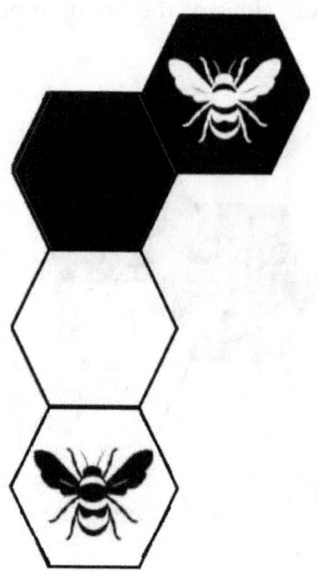

(By the way, Black should never play their Queen Bee behind their opening piece – they will find themselves on the back foot straight away and it is very hard to defend that position)

The reason the opening in the diagram above is common is that White does not want to risk that

Black might choose the C shaped opening and try to force a draw.

This opening is quite well balanced, with White having lots of attacking potential (Spiders can often be used effectively) but Black has defensive options too.

Chapter 5: More Advanced Tactics

The Pemberton Manoeuvre:

By strategically upending the nearest mug or glass of liquid, the White player can both obscure the playing field and naturally request the Black player retrieve a cloth. While the Black player is distracted the White player has plenty of time for a tactical swap.

Once the Black player has returned and mopped, the White player will now find they are in a much stronger position.

The Pemberton Manoeuvre (so-called for Coke creator and Hive *enthusiast John Pemberton who made liberal use of cola to win numerous games against his slaves in the 1860s) is a staple among the higher echelons of the* Hive *Championships.*

~ Brendan Caldwell (Shut Up and Sit Down)

The Double or Triple Pin

If an opponent has two or more powerful pieces together in a line, it is sometimes possible to pin them all with just one of your pieces:

In the diagram below, when white moves his Ant black will find that all three of his pieces are trapped.

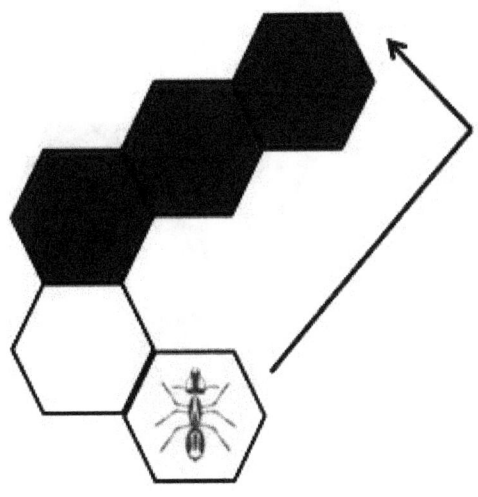

A common variation of this is to move a pin so that it traps two pieces instead of one. In the diagram below white cannot move the Beetle because it is pinned by the black Ant. Instead, he places an Ant of his own in space X, not realising that black can simply move his Ant to space Y, pinning the new Ant and continuing to pin the Beetle.

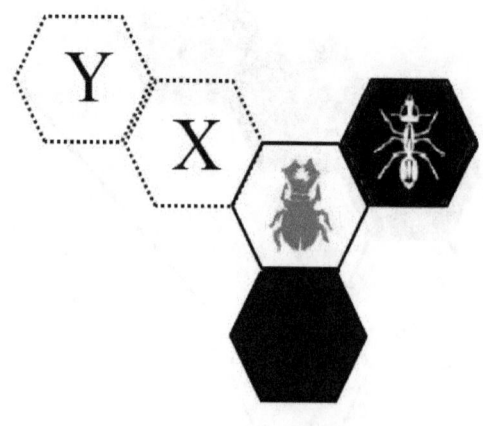

Playing Defensively

These are the key tactics used to protect the Queen Bee:

- If partially surrounded but not pinned or blocked, move the Queen Bee to a less surrounded position.
- Pinning or double-pinning opponent's attacking pieces.
- Placing a block.
- Have a friendly piece adjacent that can jump or climb out (i.e. not Ants or Spiders) if need be later.
- Position a Beetle near the Queen Bee so that it can climb on top of any opponent's Beetle that approaches.
- Use the pill bug to move a Queen Bee away from danger, or relocate attacking pieces away from the Queen Bee (more details below).

The goal of playing defensively might be to force a draw, in which case you will want to neutralise your opponent's attacks so that they cannot complete the surrounding of the Queen Bee. If you are successful, there will come a point when they cannot play any more attacks.

At this point however, you should consider whether you can go on the attack and actually win the game. I have played many games where one player went out in front early on and looked certain to win, only to run out of attacks and have to sit and watch while the other player counterattacked and won.

Releasing a Pin

If, say, your Ant is pinning an opponent's piece, you will usually find you cannot move your Ant without releasing the opponent's piece, which you do not wish to do. A solution is to move a weaker piece like a Grasshopper or a Spider to pin the opponent's piece too. That way, you can move your own Ant away on the next turn.

This manoeuvre is quite slow though, and you should only do it when you are not involved in a race.

Controlling Placement

Controlling Placement means positioning your pieces in a way that restricts where your opponent can place new pieces.

Early on in the game a player's options when deciding where to place pieces are limited. When you move or place one of your pieces, consider whether your move increases or reduces your opponent's number of options. You might be able to force them to place a piece somewhere unfavourable – somewhere it can be pinned or double-pinned perhaps, or somewhere it cannot attack from effectively.

For example, in the game below, the White pieces have been carefully positioned so that black's only option is to place a piece at position A. As soon as they do that, White will move the Spider to perform a double pin.

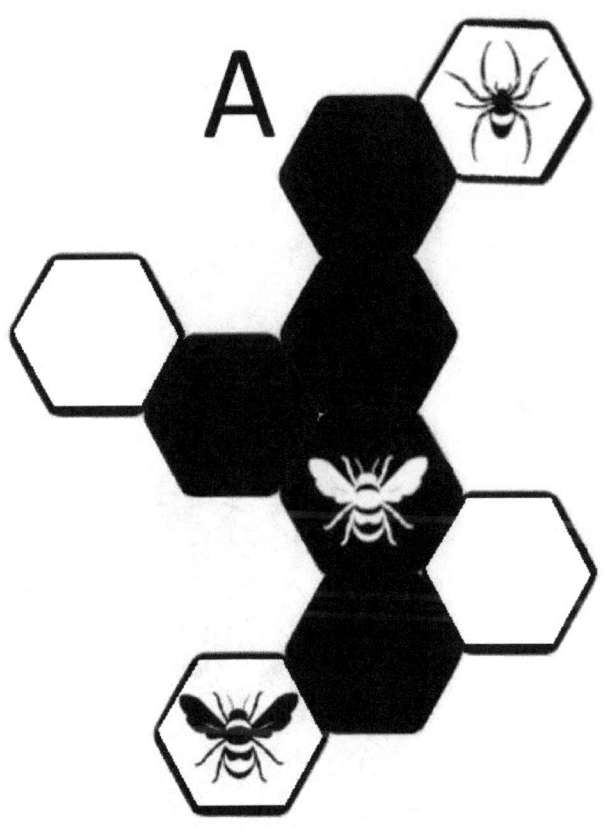

The Lockdown

The best possible outcome of a successful Controlling Placement strategy, is when your opponent is unable to move and unable to place any new pieces at all. I call this a Lockdown (but I am not sure if anyone else does).

In the game below, Black is unable to move or place a piece; White's victory will surely soon follow.

The Mosquito

If used correctly, the Mosquito can be the most powerful piece in the game.

An experienced *Hive* player will use the Mosquito tactically. They might place it next to a piece whose ability they want to adopt, or perhaps they will place it next to an Ant, then move it next turn to somewhere it can adopt a tactically useful ability.

The Mosquito can also be used to defend against Beetle attacks. If you place your Mosquito adjacent to your Queen Bee, then if an opponent's Beetle ever comes near, you should be able to put your Mosquito on top of your opponent's Beetle, neutralising it.

Also, the Mosquito and Pill Bug can form a very powerful combination. Placing them together increases the size of the area where you may move pieces around, and this can be very useful defensively, and difficult to counter.

The Ladybug

The Ladybug has the ability to climb into and out of surrounded spaces, but it also has a limited range. Within its range though, its movement pattern makes it very flexible, and later in the game it can often reach any space within a three-piece radius.

It is a very effective opening piece. Although the opening piece often gets trapped, the Ladybug has a better chance than most of eventually getting free, and if it does get out it is more likely to be able to attack the opponent's Queen Bee.

The Pill Bug

It is often placed near to its own Queen Bee so that it can rescue it later in the game, or move attackers away.

If it is to be used to rescue the Queen Bee, it will need to be placed adjacent to it. If a Queen Bee is nearly surrounded, moving it away can change the game dramatically and turn a losing position into a winning one. Note that if the Queen is pinned, the Pill Bug might actually have to help surround its own Queen Bee in order to free it so that it can be moved.

Its other defensive power is its ability to relocate attacking pieces. In this case its best position is two spaces away from the Queen Bee.

If an opponent is playing defensively and using a Pill Bug, you will have to deal with it in one of the following ways:

- Trap the Queen Bee with a Beetle so that the Pill Bug cannot move it.
- Put a Beetle on top of the Pill Bug to neutralise it.
- Surround the Pill Bug so that it has nowhere to put the pieces it has moved, and cannot use its ability.

Although the Pill Bug is primarily a defensive piece, it can also be used as an effective attacking piece if placed near to the opponent's Queen Bee. It is difficult to pin because it can move the piece that pins it a turn later. Also, it can defeat a block and place a piece in a space that is partially surrounded. If I am playing against a defensive player, I often save my Pill Bug for the final attack as it is very difficult to stop.

Playing for a Draw

Sometimes you might be happy with a draw instead of a win. Perhaps you are playing in a best-of-three (or five) match and you won the first game. Or perhaps you are just playing against a very strong opponent, and do not think that a win is a realistic possibility, especially if you are playing black.

There are three ways to get a draw in *Hive*:

1. A draw is agreed upon if neither player thinks they can win. This can happen when most of the attacking pieces on both sides are unable to move. An example of this, which is not uncommon, is when the hive contains one or more long chain of pieces, all pinning each other. This situation is not easy to engineer, but it is useful to be able to recognise it.

2. When a series of moves repeat themselves over and over, a draw is declared. For example, suppose you are about to play a very strong move, but your opponent prevents it – then you see another strong move, but your opponent prevents that too. However, suppose you now find yourself back where you started, and you are able to initiate the threat of the first strong move again, repeating the cycle. In this situation, you can force your opponent to either back down, and allow you

to make one of the strong moves, or accept a draw.

3. The game ends in a draw if both Queen Bees are surrounded simultaneously by moving a piece into the final adjacent empty space that they share. The best way to make this happen is to position your Queen Bee close to your opponent's Queen Bee early in the game, or move it closer (or even next to it) as soon as you can.

Sometimes you will find yourself trying to win a game against an opponent who is playing for the draw, and being aware of the techniques above will help you to avoid that. If you find that such an opponent has positioned their Queen Bee close to yours, and is clearly hoping to achieve a type 3 draw, one possible way out of it is to use a Forced Move, described in the section below.

The Forced Move

According to the rules of the game, players cannot 'pass'. That is, if a player <u>can</u> play a legal move, they must do so, even if the only possible move available is an undesirable one.

This means that, if you are firmly in control of the game, you might be able to restrict your opponent's options to such an extent that they are forced to make a suboptimal move, such as releasing a pin or opening a gate.

One very powerful use of the Forced Move is in avoiding an unwanted draw when the two Queen Bees are close to each other and share the last adjacent empty space. A strong player is sometimes able to shut down most or all of their opponent's possible moves, by pinning or blocking their pieces, leaving (or possibly opening up) only one possible move which the opponent <u>has</u> to play. This one possible move will be to move a piece away from being adjacent to the other player's Queen Bee, creating an empty space there. The strong player can now quickly move in, and fill the final empty space that they need to win the game.

Chapter 6: The Count

"One.
One turn.
Ah Ah Aaah!
Two.
Two Turns.
Three.
Three glorious turns!"

~ The Count (Sesame Street)

He would make a really annoying Hive *opponent.*

I want you to imagine a game between two people who play as though they are in a hurry to be somewhere else, and just play quickly without really thinking about it. They each play their Queen Bees on the second turn, and then after that each player's strategy is to alternate between placing a piece, and moving that piece into a position next to their opponent's Queen Bee. In this way, it should be possible for one player to win the game ten turns after playing their Queen Bee, as there are five empty spaces next to their opponent's Queen Bee.

Throughout the rest of this book I will use the word 'count' to mean the number of turns a player has to take before they will surround his opponent's Queen Bee.

In the example game above, who would be the winner? Well, it should always be player one, the player who went first.

In fact, at any given point in the game, the player with the lower count (or if it is a tie, the player whose turn it is) should have the advantage.

Now let us go back to that game we were imagining a few moments ago, but this time allow the two players to apply a little strategy. What should player two do to win?

Well, what they need to do is either slow player one down so that it takes them more than ten turns, or speed themselves up so they can win in fewer than ten. In other words, they need to improve (i.e. reduce) their own count, or worsen (increase) their opponent's. In practice, they will usually find it very hard to reduce their own count, so the strategy should be to increase their opponent's count.

The most common way of increasing an opponent's count is by interfering with one of their attacks, and making them take an extra move (or two) to get back into the same position.

It is important to remember though, is that if you spend one turn making your opponent waste one turn, then you have not really gained anything. Player two should be careful that in increasing player one's count, they are not increasing their own too.

There are various ways of interfering with attacks, and these were covered in the 'Playing Defensively' section of the previous chapter.

One powerful technique though, which is often overlooked by beginners, is Controlling Placement. Your opponent's success at attacking your Queen Bee might depend on being able to place a piece in a certain place. If you move or place a piece adjacent to that position, your opponent will have to place it elsewhere, potentially increasing the count (e.g. a Beetle might have further to crawl, or a Grasshopper might have to jump around a few extra times).

The beauty of using the Controlling Placement technique is that often there is no downside. It might make no difference to you whereabouts you place, say, an Ant, because on its next move it can go anywhere around the perimeter. However, if you choose the right spot, it can make a big difference to your opponent's count.

And while we are on the subject of your opponent placing pieces, I should stress that part of knowing what the count is, is knowing what unplaced pieces your opponent still has in reserve. Do not make the mistake of assuming your opponent will place a Grasshopper in a certain place, if all their Grasshoppers are on the table already. If your opponent has run out of pieces, their attack has failed.

You should get into the habit of always being aware of the count. You should know at all times whether you are ahead or not.

Many games of Hive are won and lost by a single count; being just one move ahead is often the difference between success and failure, so you must be sure <u>never</u> to waste a move.

Chapter 7: Beetlemania

The other day we were playing Hive, *when a song came on the radio by the Swedish rock band* The Hives. *We laughed. It was a nice moment.*

Half an hour later, we were playing a different board game, Ticket to Ride, *and on the radio, the DJ announced that the next song he was going to play would be by* The Beatles.

We stopped playing and looked at each other in anticipation. Could it be ...?

Then the song started. It was Eleanor Rigby.

Pity.

There is one particular attack which is very common, and very powerful indeed, and I want to talk about it at length, so I have given it a chapter of its own. I want to tell you how to play it, and I want to tell you how to defend against it if it is played against you.

The main objective of the manoeuvre is to get your Beetle on top of your opponent's Queen Bee as soon as possible. If you do that, it is likely you will be able to place several of your pieces adjacent to the Queen Bee, rather than having to place them elsewhere then move them.

There has been a lot of discussion about this manoeuvre in its various forums on the internet, and a lot of people seem to be convinced that playing this strategy will guarantee a win for the attacking player (i.e. white). This is not the case, though it will certainly give that player a big advantage if not defended against properly.

Let's look at how the attack works, shall we?

The best way of getting a Beetle on top of the Queen Bee early in the game, is by following the steps below:

1. Open with 3 pieces such as Grasshopper – Ant – Queen Bee.
2. Pin the opponent's Queen Bee with your Ant as soon as you possibly can.
3. Place a Beetle next to your Ant, in such a way that it is one space away from your opponent's Queen Bee.

At this point, the game should look something like this:

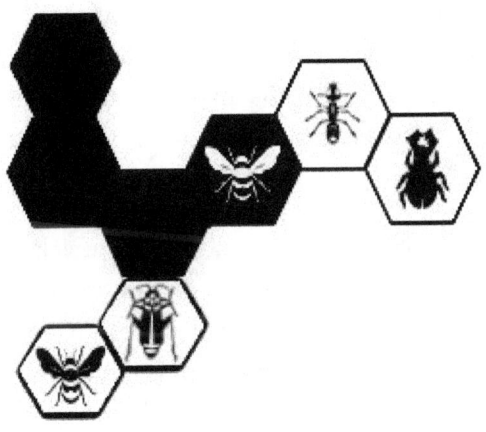

For your next few moves:

4. Move your Beetle so that it is adjacent to the opponent's Queen Bee.
5. Move your Beetle so that it is on top of the Queen Bee.
6. Place any piece (weak pieces like Spiders are fine) adjacent to the Queen Bee and your Ant.
7. Place another piece adjacent to the Queen Bee on the other side of your Ant.
8. Place another attacking piece, like an Ant, anywhere you want.
9. Move it to fill the penultimate space next to the Queen Bee. Choose the space that does not release any of your opponent's pieces.
10. Drop the Beetle from on top of the Queen into the space adjacent to your first piece (the Grasshopper). It is important to fill this space last if your opponent's first piece is a jumper or climber, and could escape.

A lot of people seem to think that this attack is special in some way. They think that it improves their count because of the way the pieces can be placed directly next to the Queen Bee. However, this is not quite the case. Even if the opponent does not defend, after pinning the Queen Bee it still takes a further 8 moves to occupy the 4 remaining spaces adjacent to the Queen Bee. In other words, two moves per space, which is standard.

Furthermore, the manoeuvre cannot win the game outright, it relies on one standard attack (moves 8 and 9 above) which can be defended against in the usual way.

However, although it is not faster than other attacks, it is more difficult to stop, especially once the Beetle gets up on top of the hive.

If you try this attack against an inexperienced player, they might not see it coming until it is too late. However, an experienced opponent probably will. If so, there are two defences they are likely to try.

Firstly, they might play a Beetle next to their own Queen Bee. He knows that you know that if you manage to get the Beetle on top of his Queen Bee, he will simply play his Beetle on top of yours, and you will not be able to play steps 6 and 7 above. If they place a Beetle therefore, you should try to pin it. This negates their threat and has the further advantage of helping you surround the Queen Bee.

Secondly, they might pin your Beetle, most likely with an Ant. If this happens, play another Beetle. If they pin that too, it is time to move onto another strategy, but you will be in a decent position, with two of your opponent's Ants unable to move for a while...

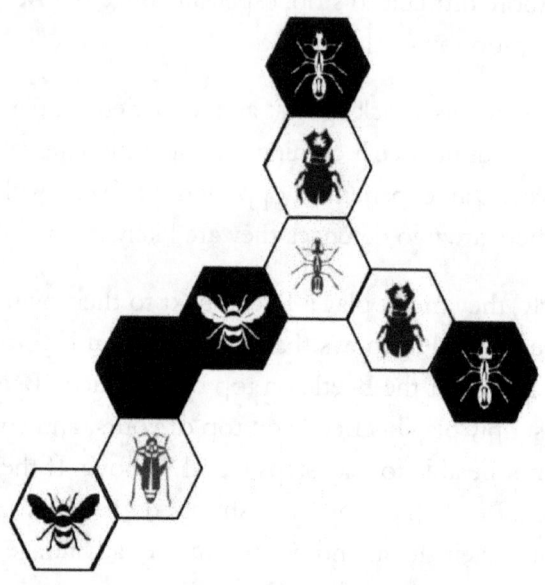

...but you should make a mental note that this particular opponent is wise to this attack, and you should try something else next time.

Chapter 8: Interview

"Thanks, Steve, for sending the book.

I like it as a quick guide for new players - it does do a good job in introducing the game and has some good elements of tactics.

I think I found a mistake, though, but I forgot what it was."

~ John Yianni

Back in chapter one, you might remember I said I had never met John Yianni, the creator of *Hive*. That is still true; however, since I wrote that, thanks to the miracle of the internet and email, I have managed to make contact with him.

I asked him for permission to use the official *Hive* artwork, which he kindly agreed to, and I also asked him for an interview, which he also kindly agreed to.

So, for the rest of this chapter, I will tell you about the on-line chat that John and I had about the history and development of the game.

I hope you find it interesting.

Firstly, I asked John the obvious question, which was where he got the idea from for the game.

"I was 18 years old," he told me, "when I first had the idea for a boardless chess-like game. I had no experience with gaming, except for games like *Monopoly* and *Risk*. But I had been designing and making board games for fun that I would play with family and friends. It was at this period that I had the idea for *Hive*, whilst watching a film on TV. The main characters of the film were two old friends that met daily in a park to play chess. They would come together, each bringing one half of the board and half the pieces."

John, unfortunately, cannot remember the name of the film. Between us, we established that it was late seventies or early eighties American film. He thinks that conflict arose when one of the men's son started dating the other's daughter, but it was a light-hearted film with an *Odd Couple* vibe.

(If this sounds familiar to anyone, please let me know, and I will pass it on. You will make John a happy man and I will mention your name here in future editions.)

"Although I don't remember the name of the film," said John, "I do remember what inspired me. Looking at the unused empty spaces on their chess board, I wondered, could I design a game that had no need of a board, so it could be very portable, but still kept the essence of what makes chess so appealing?"

John began to design the game that was in his head, and started experimenting with pieces he made from thick cardboard.

"The game was made up of squares (still thinking chess board) with symbols of Rats, Dragons, Stones, Kings, and Pigs. Some pieces moved in the same way as our current *Hive* game (the Rat had the same movement as the Ant), but then there were other pieces like the Pig that pushed other pieces around, and the Stone, which was more of a blocking piece. The object of the game was the same as *Hive*, to surround an opposing piece (in this case the King) using pieces that all moved in different ways. All the pieces needed to be connected to one another, just like *Hive*. Unlike *Hive*, though, there was a set-up position. The game played okay but not as well as I would have liked. I decided that it was not worth pursuing and so I put it away never to be played or seen again for 18 years."

Nearly two decades later, while working on a different project, John found himself drawing a hex shape on his computer.

"For some strange reason, that old game came flooding back to my mind. I kind of knew instinctively that I was onto something with the different shape of the pieces."

Once John switched from squares to hexes, the theme of using insects (perhaps because of the shape of honeycomb cells) seemed to fit better. Pieces were added and their powers defined, inspired both by chess, and the insects themselves. The Queen Bee is slow and weak, and must be defended, like the king in chess and the queen in a beehive. Grasshoppers jump over other pieces like knights, and like grasshoppers in the real world. And so on. Before long, *Hive* as we know it was born.

Although the pieces themselves, and their actions, were quickly established, the complement of pieces on each side took a while to get right. John took his prototype to a games club in Finchley, North London, and got useful feedback from the members there.

I asked John if he had any pictures of this prototype version and he sent me the picture below.

"We tried many different combinations of pieces," said John, "before we were happy with the end results. For instance, I know some people complain about the weakness of the Spider compared to the Ant and say that the Spider being the weaker piece should have a larger number than the Ants, being that the Spiders are the pawns of the *Hive* world. In some early playtests the Spiders (three) outnumbered the Ants (two), but we found we needed to have three Ants for the game to work well defensively and only really needed two Spiders."

Having heard that getting a publishing deal would be very difficult, John decided to publish the game himself. 2000-2001 was the beginning of the period that John refers to as The Homemade *Hive* Years.

"The first thousand games were made mostly by hand, from wood. I found a local cabinetmaker, who had the machinery and was willing to make the hexagonal blocks for the pieces. He would make them in batches of a few hundred at a time without finishing them. I would then take them home and set about sanding them. I also had 22000 foil hex stickers printed and one thousand printed boxes and rule booklets. The hard work was the sanding and the sticking of the stickers onto the blocks. We roped in a few good friends and family to help with the sanding and paid them with an endless supply of *KFC* and beer. My wife, Maria, and I did the rest, sticking the insect stickers onto the hexes using tweezers, and packing them in their boxes."

Getting *Hive* onto retailers' shelves was also not easy, but an early supporter was *Leisure Games*, in Finchley. The owner of that shop had heard of the game from his contacts in Finchley Games Club, and he was happy to risk taking on a few copies – these sold, and reorders followed.

Like many hopeful game designers before and since, John, with the help of Maria, decided to market the product at toy and game fairs around Europe, including the huge German game fair *Spiel*, or *Essen*.

"Our first *Essen* was in 2002, back when the fair was still small enough that you could actually breath when moving around the halls.

The day before the fair started, the venue was busy with traders and stand construction companies running here and there, all frantically trying to get their stands ready for the next day. We introduced ourselves to our neighbours who happened to be the guys from Warfrog Games. One of the guys from Warfrog, after seeing how many games we had brought with us, and trying to help prepare us for what we would expect in sales, said that most new companies only manage to sell a few games at the fair and that we should not be discouraged if we only sold around 20-30 games. He said that would be pretty good going for our first time. That's not what we wanted to hear considering we had brought 500 games with us, but what could we do? We just got on with the wallpapering of the stand and hanging up the very large (and expensive) posters we had had printed.

"The first day of the fair started with a bang. As the doors opened there was a rush of guys running into the halls with intent, clutching lists and pulling along trolleys, ready to buy the new games on sale at the show. It seemed that these list-clutching trolley pullers had already decided which games they would buy beforehand (and Warfrog was doing a booming trade) but unfortunately *Hive* did not seem to be on their lists. Apart from a few people who had sat down to play, our booth was not doing a great job at attracting customers. I was now suspecting that the guy from Warfrog had been a tad optimistic with his estimation of 20-30 games.

"But slowly and surely, we started to notice a pattern. The same faces, the same people coming back to our booth bringing along other people in tow, to try the game.

"Soon enough our booth was packed with people playing *Hive*, the tables were full, and any available floor space had quickly filled up. We had to step over people to get to the games that where all piled up at the back of the stand."

After that first day, John and Maria were …ahem … buzzing.

"We never expected it to be so easy. The rest of the weekend was even more of a surprise, with thousands of people packing the halls. If you stepped out of your booth you could be swept along by the crowds, and had to fight your way back. Along with the public we also had a German games publisher who visited our stand a few times, showing the game to people he had brought along with him. Eventually he sat me down and introduced himself; it turned out that he was one of the founders of the big German publishers' houses. He was very keen to sign a publishing deal with me for *Hive*, but because I was very new to the games business, I decided to decline his offer and go it alone."

Over the years that same publisher became a frequent visitor to John's stand in Essen, and eventually they came to an arrangement to allow the company to distribute *Hive* in Germany. They are still the German distributors to this day.

"We left our first Essen games fair with high expectations about the future of *Hive*. With offers from publishers (we came home to another offer from Kosmos). *Hive* was voted the best game in show on the Fairplay public poll that first year, and with almost all of our 500 games sold at the fair. Without doubt our first Essen was a massive success for us."

Before long, the homemade wooden pieces were upgraded to the professionally-made resin pieces that we know and love today. These wonderfully tactile pieces are made using a high-pressured heat mould, then hand painted, and finally 'tumbled' to smooth and polish them.

New editions were created, and new pieces were added too – first the Mosquito, then the Ladybug, then the Pillbug. Originally, new pieces were given away at trade fairs to promote the game, but eventually, due to popular demand, they were made available to buy.

So how big is *Hive* now, in 2020?

"I don't really know how many players play the game on a regular basis, but I can tell you that the game is distributed in about 35 countries. It has sold around 700,000 copies since its humble beginnings, and sales grow exponentially year by year. There are small and large tournaments played annually around the world, and some of the larger tournaments have about 3000 players."

I don't know about you, but one thing I have often wondered about with strategy games, is how good the people who invent them are. So, I asked John to tell me how good he is at *Hive*.

"I am not the world's greatest. I spend most of my time demonstrating the game to new players; I don't really get time to play with high ranking players, so I don't get to improve my game much. I'm amazed at how some of those top players have taken my humble game and elevated it to a level I would never have dreamed of in those early years."

As well as *Hive*, John has designed a number of other strategy games, including *Army of Frogs*, *Logan Stones*, and *Junkyard Races*. The one that caught my eye though, was *Tatsu*. In this game, players roll dice to move pieces that represent dragons around a circular board. It seems to me that it has a lot in common with another classic board game.

"Yes," John confirmed, "*Tatsu* was inspired by its grandfather Backgammon. *Tatsu* is to Backgammon as *Hive* is to Chess."

"Being myself of Greek origin, I was very much accustomed to having a Backgammon set present at most family gatherings. There are a many different variations played with a Backgammon set, but the standard game is where you start with all your stones in set positions and move them around the board and try to bear them all off. In this Standard game you can send back any of your opponent's stones that are not defended, and this was the inspiration behind the Water Dragon in *Tatsu*.

"The second variation is called Blocking Backgammon. This is easier to set up as all the stones start off the board, and is often the game taught to young children because it's easier to pick up and play. As the name entails this game is played by blocking your opponents undefended stones as you try to bear off all of yours. This was the inspiration behind the Vine Dragon in *Tatsu*.

"There are some similarities between Backgammon and *Tatsu* but there are quite a few differences too. *Tatsu* has the same number of spaces around the board as Backgammon, so the dice probabilities will feel quite familiar to a Backgammon player. With *Tatsu* you are limited to only two stones per segment, but with Backgammon you are able to have unlimited stones in one segment.

"The other differences in the games are the winning conditions. In *Tatsu*, instead of it being a race to see who can bear off all their stones first, there are two winning conditions: you can ether win by expelling all your opponent's stones from play, or by eliminating all of one type of your opponent's stones by landing on them with your Fire Dragons. Whereas Backgammon has only one type of stone, *Tatsu* has three, and with fewer numbers of the more powerful stones. This gives the game a really elegant balance, as you need the more powerful stones to win by elimination, but these are vulnerable to attack and could lose you the game very quickly if left unprotected.

"I personally love playing *Tatsu*, it's a game that is easy to pick up and play with just about anyone and because you are only making decisions based on the dice roll in that particular moment. It can also be played in a quite relaxing atmosphere, over a cool drink, on a lovely sunny day.

So, apart from *Hive* and *Tatsu*, what other games does John enjoy playing?

"I'd much prefer to play a quick family friendly game than a long, complicated game. So, games such as *King Of Tokyo*, *Camel Up*, *Dixit* and *Loony Quest* will see a lot of table time."

John did add however, that there is one longer game he does enjoy - *Xia: Legends Of A Drift System*.

"I love its sandbox game play," he said. "It reminds me of a video game that I spent too much time playing as a kid, *Elite* for the Commodore 64."

Finally, I asked John what the future holds for *Hive*. There are no plans to add any more pieces to the game any time soon, but with the 20th anniversary of the first sale coming up, he does have plans to create some cool stuff to celebrate this.

"The plan is to make a limited run of the game using Metal game pieces, with enamelled insect designs, inside a hexagonal PU leather zip-up case. These will be a limited number (still to be determined) and will be just for the 20th Anniversary of the game.

"We are also launching a hand-made hexagonal, beautiful wooden case for *Hive Pocket*. Though these are not going to be a limited run, but because they are hand made here in the UK, they will only be produced in small numbers.

"And thirdly, we are in negotiations with a digital studio about a new iOS game which we hope will launch in time for the celebration of *Hive*'s 20th year."

Chapter 9: Final Thoughts

Many believe Hive *is on the verge of going mainstream. One indicator of this is the (perhaps surprising) fact that the singer Nicki Minaj is a celebrity* Hive *player. Her 2012 song* Beez in the Trap *is a tribute to the game.*

At least, I think it is; she has not returned my emails asking her to confirm this. However, what other possible reason could there be for trapping bees? If I had written the song I would have used the word "pin" rather than "trap" (and I would also have spelt "bees" correctly). Listening to the lyrics it is difficult to see exactly what she is getting at, but she uses the word mothercluckers a lot, which I believe is a common euphemism for Grasshopper, and towards the end of the song, she talks about defecating over a large area, which is a well-known variation on the Pemberton manoeuvre.

Well, that just about brings us to the end of this book. I hope you have enjoyed reading it as much as I have enjoyed researching and writing it.

If ever you want to read even more about strategy and game theory, you could check out Randy Ingersoll's book *Play Hive like a Champion* or Joe Schultz's *The Canon Of Hive*. Both of these are outsize paperbacks (and much thicker than this book too) with lots of discussion around the different techniques, and lots of diagrams taken from games.

However, the most important thing for you to do now that you know the rules and the key strategies is to practice. The more you play the more confident you will become with counting and planning ahead.

Hive is a great lunchtime game if you are at school or work, as you can play it just about anywhere. All you need is an opponent. I played most of my early games against a friend in the staff canteen at work, and curious colleagues were forever coming over to see what we were playing. Most of them are now *Hive* players too.

I am usually available to play at a moment's notice, as I carry my copy of *Hive Pocket* with me at all times; I can confirm that it does indeed fit in a pocket.

(And yes, that means I have ants in my pants.)

If you cannot find an opponent though, there are several online versions of the game. I hesitate to name any of them because technology is changing all the time, and apps and websites come and go. Ratehr than recommend something that will be obsolete in a few years, I am tempted just to advise you to enter 'Play *Hive* Online' into Google (I think I am safe assuming Google will not be forgotten any time soon).

I will just mention BoardgameArena.com though, as that is a great site that I have been using for years. It connects you to thousands of *Hive* players all over the world, of every level of expertise, and has a lovely interface that feels almost like playing the game live. It has lots of other games on it too, some of which (like *Carcassonne*, *Terra Mystica*, and *Seven Wonders*) you have to be a paying member to use; however, if you just want to play *Hive*, it's free.

Don't worry about winning or losing at first – winning is nice, but one of the best ways to learn is to lose a game to somebody with more experience.

If you prefer to play on your phone, there is also an app.

I should warn you though, …

[brace yourself]

…the software does have bugs in it!

On that terrible joke, I am going to sign off now. I hope you have enjoyed this little book and I hope it improves the way you play the game.

Happy Hiving.

Steve.

Thank you for reading this book.

If you enjoyed it, you might also like …

Ticket to Carcassonne is your introduction to the many, many games in the board game hobby, and is available now as a paperback or an e-book, from all good online retailers named after South American Rivers.

www.ingramcontent.com/pod-product-compliance
Lightning Source LLC
Chambersburg PA
CBHW050244220526
45465CB00002B/538